SOME FAMOUS PROBLEMS

of the

THEORY OF NUMBERS

and in particular

Waring's Problem

An Inaugural Lecture delivered before the
University of Oxford

BY

G. H. HARDY, M.A., F.R.S.

Fellow of New College
Savilian Professor of Geometry in the University of Oxford
and late Fellow of Trinity College, Cambridge

SOME FAMOUS PROBLEMS OF THE THEORY OF NUMBERS.

IT is expected that a professor who delivers an inaugural lecture should choose a subject of wider interest than those which he expounds to his ordinary classes. This custom is entirely reasonable; but it leaves a pure mathematician faced by a very awkward dilemma. There are subjects in which only what is trivial is easily and generally comprehensible. Pure mathematics, I am afraid, is one of them; indeed it is more: it is perhaps the one subject in the world of which it is true, not only that it is genuinely difficult to understand, not only that no one is ashamed of inability to understand it, but even that most men are more ready to exaggerate than to dissemble their lack of understanding.

There is one method of meeting such a situation which is sometimes adopted with considerable success. The lecturer may set out to justify his existence by enlarging upon the overwhelming importance, both to his University and to the community in general, of the particular studies on which he is engaged. He may point out how ridiculously inadequate is the recognition at present afforded to them; how urgent it is in the national interest that they should be largely and immediately re-endowed; and how immensely all of us would benefit were we to entrust him and his colleagues with a predominant voice in all questions of educational administration. I have observed friends of my own, promoted to chairs of various subjects in various Universities, addressing themselves to this task with an eloquence and courage which it would be impertinent in me to praise. For my own part, I trust that I am not lacking in respect either for my subject or myself. But, if I am asked to explain how, and why, the solution of the problems which occupy the best energies of my life is of importance in the general life of the community, I must decline the unequal contest: I have not the effrontery to develop a thesis so palpably untrue. I must leave it to the engineers and the chemists to expound, with justly prophetic fervour, the benefits conferred on civilization by gas-engines, oil, and explosives. If I could attain every scientific ambition of my life, the frontiers of the Empire would not be advanced, not even a black man would be blown to pieces, no one's fortune would be made, and least of all my own. A pure mathematician must leave to happier colleagues the great task of alleviating the sufferings of humanity.

I suppose that every mathematician is sometimes depressed, as certainly I often am myself, by this feeling of helplessness and futility. I do not profess to have any very satisfactory consolation to offer. It is possible that the life of a mathematician is one which no perfectly reasonable man would elect to live. There are, however, one or two reflections from which I have sometimes found it possible to extract a certain amount of comfort. In the first place, the study of mathematics is, if an unprofitable, a perfectly harmless and innocent occupation, and we have learnt that it is something to be able to say that at any rate we do no harm. Secondly, the scale of the universe is large, and, if we are wasting our time, the waste of the lives of a few university dons is no such overwhelming catastrophe. Thirdly, what we do may be small, but it has a certain character of permanence; and to have produced anything of the slightest permanent interest, whether it be a copy of verses or a geometrical theorem, is to have done something utterly beyond the powers of the vast majority of men. And, finally, the history of our subject does seem to show conclusively that it is no such mean study after all. The mathematicians of the past have not been neglected or despised; they have been rewarded in a manner, undiscriminating perhaps, but certainly not ungenerous. At all events we can claim that, if we are foolish in the object of our devotion, we are only in our small way aping the folly of a long line of famous men, and that, in these days of conflict between ancient and modern studies, there must surely be something to be said for a study which did not begin with Pythagoras, and will not end with Einstein, but is the oldest and the youngest of all.

It seemed to me for a moment, when I was considering what subject I should choose, that there was perhaps one which might, in a philosophic University like this, be of wider interest than ordinary technical mathematics. If modern pure mathematics has any important applications, they are the applications to philosophy made by the mathematical logicians of the last thirty years. In the sphere of philosophy we mathematicians put forward a strictly limited but absolutely definite claim. We do not claim that we hold in our hands the key to all the riddles of existence, or that our mathematics gives us a vision of reality to which the less fortunate philosopher cannot attain; but we do claim that there are a number of puzzles, of an abstract and elusive kind, with which the philosophers of the past have struggled ineffectually, and of which we now can give a quite definite and explicit solution. There is a certain region of philosophical territory which it is our intention to annex. It is a strictly demarcated region, but it has suffered under the misrule of philosophers for generations, and it is ours by right; we propose to accept the mandate for it, and to offer it the opportunity of self-determination under the mathematical flag. Such at any rate is the thesis which I hope it may before long be my privilege to defend.

It seemed to me, however, when I considered the matter further, that there are two fatal objections to mathematical philosophy as a subject for an inaugural address. In the first place the subject is one which requires a certain amount of application and preliminary study. It is not that it is a subject, now that the foundations have been laid, of any extraordinary difficulty or obscurity; nor that it demands any wide knowledge of ordinary mathematics. But there are certain things that it does demand; a little thought and patience, a little respect for mathematics, and a little of the mathematical habit of mind which comes fully only after long years spent in the company of mathematical ideas. Something, in short, may be learnt in a term, but hardly in a casual hour.

In the second place, I think that a professor should choose, for his inaugural lecture, a subject, if such a subject exists, to which he has made himself some contribution of substance and about which he has something new to say. And about mathematical philosophy I have nothing new to say; I can only repeat what has been said by the men, Cantor and Frege in Germany, Peano in Italy, Russell and Whitehead in England, who have originated the subject and moulded it now into something like a definite form. It would be an insult to my new University to offer it a watered synopsis of some one else's work. I have therefore finally decided, after much hesitation, to take a subject which is quite frankly mathematical, and to give a summary account of the results of some researches which, whether or no they contain anything of any interest or importance, have at any rate the merit that they represent the best that I can do.

My own favourite subject has certain redeeming advantages. It is a subject, in the first place, in which a large proportion of the most remarkable results are by no means beyond popular comprehension. There is nothing in the least popular about its *methods*; as to its votaries it is the most beautiful, so by common consent it is the most difficult of all branches of a difficult science; but many of the actual results are such as can be stated in a simple and striking form. The subject has also a considerable historical connexion with this particular chair. I do not wish to exaggerate this connexion. It must be admitted that the contributions of English mathematicians to the Theory of Numbers have been, in the aggregate, comparatively slight. Fermat was not an Englishman, nor Euler, nor Gauss, nor Dirichlet, nor Riemann; and it is not Oxford or Cambridge, but Göttingen, that is the centre of arithmetical research to-day. Still, there has been an English connexion, and it has been for the most part a connexion with Oxford and with the Savilian chair.

The connexion of Oxford with the theory of numbers is in the main a nineteenth-century connexion, and centres naturally in the names of Sylvester and Henry Smith. There is a more ancient, if indirect, connexion which I ought not altogether to forget. The theory of numbers, more than any other branch of pure mathematics, has begun by being an empirical science. Its most famous theorems have all been conjectured, sometimes a hundred years or more before they have been proved; and they have been suggested by the evidence of a mass of computation. Even now there is a considerable part to be played by the computer; and a man who has to undertake laborious arithmetical computations is hardly likely to forget what he owes to Briggs. However, this is ancient history, and it is with Sylvester and Smith that I am concerned to-day, and more particularly with Smith.

Henry Smith was very many things, but above all things a most brilliant arithmetician. Three-quarters of the first volume of his memoirs is occupied with the theory of numbers, and Dr. Glaisher, his mathematical biographer, has observed very justly that, even when he is primarily concerned with other matters, the most striking feature of his work is the strongly arithmetical spirit which pervades the whole. His most remarkable contributions to the theory are contained in his memoirs on the arithmetical theory of forms, and in particular in the famous memoir on the representation of numbers by sums of five squares, crowned by the Paris Academy and published only after his death. This memoir is peculiarly interesting to me, for the problem is precisely one of those of which I propose to speak to-day; and I may perhaps add one comment on the surprising history set out in Dr. Glaisher's introduction. The name of Minkowski is familiar to-day to many, even in Oxford, who have certainly never read a line of Smith. It is curious to contemplate at a distance the storm of indignation which convulsed the mathematical circles of England when Smith, bracketed after his death with the then unknown German mathematician, received a greater honour than any that had been paid to him in life.

The particular problems with which I am concerned belong to what is called the 'additive' side of higher arithmetic. The general problem may be stated as follows.

Suppose that n is any positive integer, and

$$\alpha_1, \ \alpha_2, \ \alpha_3, \ \ldots$$

positive integers of some special kind, squares, for example, or cubes, or perfect kth powers, or primes. We consider all possible expressions of n in the form

$$n = \alpha_1 + \alpha_2 + \cdots + \alpha_s,$$

where s may be fixed or unrestricted, the α's may or may not be necessarily distinct, and order may or may not be relevant, according to the particular problem on which we are engaged. We denote by

$$r(n)$$

the number of representations which satisfy the conditions of the problem. Then *what can we say about $r(n)$?* Can we find an exact formula for $r(n)$, or an approximate formula valid for large values of n? In particular, is $r(n)$ *always positive?* Is it always possible, that is to say, to find at least *one* representation of n of the type required? Or, if this is not so, is it at any rate always possible when n is sufficiently large?

I can illustrate the nature of the general problem most simply by considering for a moment an entirely trivial case. Let us suppose that there are three different α's only, viz. the numbers 1, 2, 3; that repetitions of the same α are permissible; that the order of the α's is irrelevant; and that s, the number of the α's, is unrestricted. Then it is easy to see that $r(n)$, the number of representations, is the number of solutions of the equation

$$n = x + 2y + 3z$$

in positive integers, including zero.

There are various ways of solving this extremely simple problem. The most interesting for our present purpose is that which rests on an analytical foundation, and uses the idea of the *generating function*

$$f(x) = 1 + \sum_{1}^{\infty} r(n)x^n,$$

in which the coefficients are the values of the arithmetical function $r(n)$. It follows immediately from the definition of $r(n)$ that

$$f(x) = (1 + x + x^2 + \ldots)(1 + x^2 + x^4 + \ldots)(1 + x^3 + x^6 + \ldots)$$
$$= \frac{1}{(1-x)(1-x^2)(1-x^3)};$$

and, in order to determine the coefficients in the expansion, nothing more than a little elementary algebra is required. We find, by the ordinary theory of partial fractions, that

$$f(x) = \frac{1}{6(1-x)^3} + \frac{1}{4(1-x)^2} + \frac{17}{72(1-x)} + \frac{1}{8(1+x)}$$
$$+ \frac{1}{9(1-\omega x)} + \frac{1}{9(1-\omega^2 x)},$$

where ω and ω^2 denote as usual the two complex cube roots of unity. Expanding the fractions, and picking out the coefficient of x^n, we obtain

$$r(n) = \frac{(n+3)^2}{12} - \frac{7}{72} + \frac{(-1)^n}{8} + \frac{2}{9}\cos\frac{2n\pi}{3}.$$

It is easily verified that the sum of the last three terms can never be as great as $\frac{1}{2}$, so that $r(n)$ is the integer nearest to

$$\frac{(n+3)^2}{12}.$$

The problem is, as I said, quite trivial, but it is interesting none the less. A great deal of work has been done on problems similar in kind, though naturally far more complex and difficult in detail, by Cayley and Sylvester, for example, in the last century, and by Glaisher, and above all by MacMahon, in this. And even this problem, simple as it is, has sufficient content to bring out clearly certain principles of cardinal importance.

In particular, the solution of the problem shows quite clearly that, if we are to attack these 'additive' problems by analytic methods, it is in the theory of integral power series

$$\sum a_n x^n$$

that the necessary machinery must be found. It is this character-istic which distinguishes this theory sharply from the other great side of the analytic theory of numbers, the 'multiplicative' theory, in which the fundamental idea is that of the resolution of a number into primes. In the latter theory the right weapon is generally not a power series, but what is called a Dirichlet's series, a series of the type

$$\sum a_n n^{-s}.$$

It is easy to see this by considering a simple example. One of the most interesting functions of the multiplicative theory is $d(n)$, the number of divisors of n. The associated power series

$$\sum d(n) x^n$$

is easily transformed into the series

$$\sum \frac{x^n}{1-x^n},$$

called Lambert's series. The function is an interesting one, but some-what unmanageable, and certainly not one of the fundamental functions of analysis. The corresponding Dirichlet's series is far more fundamental; it is in fact

$$\sum \frac{d(n)}{n^s} = \left(\sum \frac{1}{n^s}\right)^2 = \left(\zeta(s)\right)^2,$$

the square of the famous Zeta function of Riemann.

The underlying reason for this distinction is fairly obvious. It is natural to *multiply* primes and unnatural to *add* them. Now

$$m^{-s} \times n^{-s} = (mn)^{-s},$$

so that, in the theory of Dirichlet's series, the terms combine naturally with one another in a 'multiplicative' manner. But

$$x^m \times x^n = x^{m+n},$$

so that the multiplication of two terms of a power series involves an additive operation on their ranks. It is thus that the Dirichlet's series rather than the power series proves to be the proper weapon in the theory of primes.

It would be difficult for anybody to be more profoundly interested in anything than I am in the theory of primes; but it is not of this theory that I propose to speak to-day, and we must return to our general additive problem. As soon as we try to specialize the problem in some more interesting manner, two problems stand out as calling for research. Each of them, naturally, is only the representative of a class.

The first of these problems is the *problem of partitions*. Let us suppose now that the α's are *any* positive integers, and that (as in the trivial problem) repetitions are allowed, order is irrelevant, and s is unrestricted. The problem is then that of expressing n in any manner as a sum of integral parts, or of solving the equation

$$n = x + 2y + 3z + 4u + 5v + \ldots,$$

and $r(n)$ or, as it is now more naturally written, $p(n)$, is the number of *unrestricted partitions* of n. Thus

$$5 = 1 + 1 + 1 + 1 + 1 = 1 + 1 + 1 + 2$$
$$= 1 + 2 + 2 = 1 + 1 + 3 = 2 + 3 = 1 + 4 = 5,$$

so that $p(5) = 7$. The generating function in this case was found by Euler, and is

$$f(x) = 1 + \sum_1^\infty p(n)x^n = \frac{1}{(1-x)(1-x^2)(1-x^3)\ldots}.$$

I do not wish to discuss this problem in any detail now, but the form of the generating function calls for one or two general remarks. In the trivial problem the generating function was *rational*, with a finite number of poles all situated upon the unit circle. Here also we are led to a power series, or infinite product, convergent inside the unit circle; but there the resemblance ends. This function will be recognized by any one familiar with the theory of elliptic functions; it is an elliptic modular function; and, like all such functions, it has the unit circle as a continuous line of singularities and does not exist at all outside. It is easy to imagine the immensely increased difficulties of any analytic solution of the problem.

It was conjectured by a very brilliant Hungarian mathematician, Mr. G. Pólya, five or six years ago, that *any* function represented by a power series whose coefficients are *integers*, and which is convergent inside the unit circle, must behave, in this respect, like one or other of the two generating functions which we have considered. Either such a function is a rational function, that is to say, completely elementary; or else the unit circle is a line of essential singularities. I believe that a proof of this theorem has now been found by Mr. F. Carlson of Upsala, and is to be published shortly in the *Mathematische Zeitschrift*. It is difficult for me to give reasoned praise to a memoir which I have not seen, but I can recommend the theorem to your attention with confidence as one of the most beautiful of recent years.

The problem of partitions is one to which, in collaboration with the Indian mathematician, Mr. S. Ramanujan, I have myself devoted a great deal of work. The principal result of our work has been the discovery of an approximate formula for $p(n)$ in which the leading term is

$$\frac{1}{2\pi\sqrt{2}}\frac{d}{dn}\frac{e^{\frac{2\pi}{\sqrt{6}}\sqrt{n-\frac{1}{24}}}}{\sqrt{n-\frac{1}{24}}},$$

and which enables us to approximate to $p(n)$ with an accuracy which is almost uncanny. We are able, for example, by using 8 terms of our formula, to calculate $p(200)$, a number of 13 figures, with an error of .004. I have set out the details of the calculation in Table I. The value of $p(200)$ was subsequently verified by Major MacMahon, by a direct computation which occupied over a month.

The formulae connected with this problem are very elaborate, and except on the purely numerical side, where the results of the theory are compared with those of computation, it is not very well suited for a hasty exposition; and I therefore pass on at once to the principal object of my lecture, the very famous problem known, after a Cambridge professor of the eighteenth century, as *Waring's Problem*.

TABLE I.

$$p(200)$$

	$3,972,998,993,185.896$
	$36,282.978$
$-$	87.555
$+$	5.147
$+$	1.424
$+$	0.071
$+$	0.000
$+$	0.043
	$3,972,999,029,388.004$

We suppose now that every α is a perfect k-th power m^k, k being fixed in each case of the problem which we consider; m may be of either sign if k is even, but must be positive if k is odd. In either case we allow m to be zero. Repetitions are permitted, as in our previous problems; but it is more convenient now to take account of the order of the α's; and s, which was formerly unrestricted, is now fixed in each case of the problem, like k. The problem is therefore that of determining the number of representations of a number n as the sum of s positive k-th powers. Thus Henry Smith's problem, the problem of five squares, is the particular case of Waring's problem in which k is 2 and s is 5. The problem has a long history, which centres round this simplest case of squares; a history which began, I suppose, with the right-angled triangles of Pythagoras, and has been continued by a long succession of mathematicians, including Fermat, Euler, Lagrange, and Jacobi, down to the present day. I will begin by a summary of what is known in the simplest case, where the solution is practically complete.

A number n is the sum of **two** squares if and only if it is of the form

$$n = M^2 P,$$

where P is a product of primes, all different and all of the form $4k+1$. In particular, a prime number of the form $4k + 1$ can be expressed as the sum of two squares, and substantially in only one way. Thus $5 = 1^2 + 2^2$, and there is no other solution except the solutions $(\pm 1)^2 + (\pm 2)^2$, $(\pm 2)^2 + (\pm 1)^2$, which are not essentially different, although it is convenient to count them as distinct. The number of numbers less than x, and expressible as the sum of two squares, is approximately

$$\frac{Cx}{\sqrt{\log x}},$$

where C is a certain constant. The last result was proved by Landau in 1908; all the rest belong to the classical theory.

A number is the sum of **three** squares unless it is of the form

$$4^\alpha (8k + 7),$$

when it is not so expressible. *Every* number may be expressed by **four** squares, and *a fortiori* by five or more. It is this last theorem of Lagrange that I would ask you particularly to bear in mind.

If s, the number of squares, is even and less than 10, the number of representations may be expressed in a very simple form by means of the divisors of n. Thus the number of representations by 4 squares, when n is odd, is 8 times the sum of the divisors of n; when n is even, it is 24 times the sum of the odd divisors; and there are similar results for 2 squares, or 6, or 8.

When s is 3, 5, or 7, the number of representations can also be found in a simple form, though one of a very different character.

Suppose, for example, that s is 3. The problem is in this case essentially the same as that of determining the number of classes of binary arithmetical forms of determinant $-n$; and the solution depends on certain finite sums of the form

$$\sum \beta, \quad \sum \gamma,$$

extended over quadratic residues β or non-residues γ of n.

When s, whether even or odd, is greater than 8, the solution is decidedly more difficult, and it is only very recently that a uniform method of solution, for which I must refer you to some recent memoirs of Mr. L. J. Mordell and myself, has been discovered. For the moment I wish to concentrate your attention on two points: the first, that *an expression by* 4 *squares is always possible, while one by* 3 *is not*; and the second, that the existence of numbers not expressible by 3 squares is revealed at once by the quite trivial observation that no number so expressible can be congruent to 7 to modulus 8.

It is plain, when we proceed to the general case, that any number n can be expressed as a sum of k-th powers; we have only to take, for example, the sum of n ones. And, when n is given, there is a *minimum* number of k-th powers in terms of which n can be expressed; thus

$$23 = 2 \cdot 2^3 + 7 \cdot 1^3$$

is the sum of 9 cubes and of no smaller number. But it is not at all plain (and this is the point) that this minimum number cannot tend to infinity with n. It does not when $k = 2$; for then it cannot exceed 4. And Waring's Problem (in the restricted sense in which the name has commonly been used) is the problem of proving that the minimum number is similarly bounded in the general case. It is not an easy problem; its difficulty may be judged from the fact that it took 127 years to solve.

We may state the problem more formally as follows. Let k be given. Then there may or may not exist a number m, the same for all values of n, and such that n can always be expressed as the sum of m k-th powers or less. If any number m possesses this property, all larger numbers plainly possess it too; and among these numbers we may select the *least*. This least number, which will plainly depend on k, we call $g(k)$; thus $g(k)$ is, by definition, the least number, if such a number exists, for which it is true that

> *'every number is the sum of $g(k)$ k-th powers or less'*.

We have seen already that $g(2)$ exists and has the value 4.

In the third edition of his *Meditationes Algebraicae*, published in Cambridge in 1782, Waring asserted that every number is the sum of not more than 4 squares, not more than 9 cubes, not more than 19 fourth powers, *et sic deinceps*. A little more precision would perhaps have been desirable; but it has generally been held, and I do not question that it is true, that what Waring is asserting is precisely the existence of $g(k)$. He implies, moreover, that $g(2) = 4$ and $g(3) = 9$; and both of these assertions are correct, though in the first he had been anticipated by Lagrange. Whether $g(4)$ is or is not equal to 19 is not known to-day.

Waring advanced no argument of any kind in support of his assertion, and it is in the highest degree unlikely that he was in possession of any sort of proof. I have no desire to detract from the reputation of a man who was a very good mathematician if not a great one, and who held a very honourable position in a University which not even Oxford has persuaded me entirely to forget. But there is a tendency to exaggerate the profundity implied by the enunciation of a theorem of this particular kind. We have seen this even in the case of Fermat, a mathematician of a class to which Waring had not the slightest pretensions to belong, whose notorious assertion concerning 'Mersenne's numbers' has been exploded, after the lapse of over

250 years, by the calculations of the American computer Mr. Powers. No very laborious computations would be necessary to lead Waring to a highly plausible speculation, which is all I take his contribution to the theory to be; and in the Theory of Numbers it is singularly easy to speculate, though often terribly difficult to prove; and it is only proof that counts.

The next advance towards the solution of the problem was made by Liouville, who established the existence of $g(4)$. Liouville's proof, which was first published in 1859, is quite simple and, as the simplest example of an important type of argument, is worth reproducing here. It may be verified immediately that

$$6X^2 = 6(x^2 + y^2 + z^2 + t^2)^2$$
$$= (x + y)^4 + (x - y)^4 + (z + t)^4 + (z - t)^4$$
$$+ (x + z)^4 + (x - z)^4 + (t + y)^4 + (t - y)^4$$
$$+ (x + t)^4 + (x - t)^4 + (y + z)^4 + (y - z)^4;$$

and since, by Lagrange's theorem, any number X is the sum of 4 squares, it follows that any number of the form $6X^2$ is the sum of 12 biquadrates. Hence any number of the form $6(X^2 + Y^2 + Z^2 + T^2)$ or, what is the same thing, any number of the form $6m$, is the sum of 48 biquadrates. But *any* number n is of the form $6m + r$, where r is 0, 1, 2, 3, 4, or 5. And therefore n is, at worst, the sum of 53 biquadrates. That is to say, $g(4)$ exists, and does not exceed 53. Subsequent investigators, refining upon this argument, have been able to reduce this number to 37; the final proof that $g(4) \leq 37$, the most that is known at present, was given by Wieferich in 1909. The number

$$79 = 4 \cdot 2^4 + 15 \cdot 1^4$$

needs 19 biquadrates, and no number is known which needs more. There is therefore still a wide margin of uncertainty as to the actual value of $g(4)$.

The case of cubes is a little more difficult, and the existence of $g(3)$ was not established until 1895, when Maillet proved that $g(3) \leqq 17$. The proof then given by Maillet rests upon the identity

$$6x(x^2 + y^2 + z^2 + t^2)$$
$$= (x+y)^3 + (x-y)^3 + (x+z)^3 + (x-z)^3 + (x+t)^3 + (x-t)^3,$$

and the known results concerning the expression of a number by 3 squares. It has not the striking simplicity of Liouville's proof; but it has enabled successive investigators to reduce the number of cubes, until finally Wieferich, in 1909, proved that $g(3) \leqq 9$. As 23 and 239 require 9 cubes, the value of $g(3)$ is in fact exactly 9. It is only for $k = 2$ and $k = 3$ that the actual value of $g(k)$ has been determined. But similar existence proofs were found, and upper bounds for $g(k)$ determined, by various writers, in the cases $k = 5$, 6, 7, 8, and 10.

Before leaving the problem of the cubes I must call your attention to another very beautiful theorem of a slightly different character. The numbers 23 and 239 require 9 cubes, and it appears, from the results of a survey of all numbers up to 40,000, that no other number requires so many. It is true that this has not actually been proved; but it *has* been proved (and this is of course the essential point) that the number of numbers which require as many cubes as 9 is *finite*.

This singularly beautiful theorem, which is due to my friend Professor Landau of Göttingen, and is to me as fascinating as anything in the theory, also dates from 1909, a year which stands out for many reasons in the history of the problem. It is of exceptional interest not only in itself but also on account of the method by which it was proved, which utilizes some of the deepest results in the modern theory of the asymptotic distribution of primes, and made it, until very recently, the only theorem of its kind erected upon a genuinely transcendental foundation. To me it has a personal interest also, as being the only theorem of the kind which, up to the present, defeats the new analytic method by which Mr. Littlewood and I have recently studied the problem.

Landau's theorem suggests the introduction of another function of k, which I will call $G(k)$, of the same general character as $g(k)$, but I think probably more fundamental. This number $G(k)$ is defined as being the least number for which it is true that

'*every number* FROM A CERTAIN POINT ONWARDS *is the sum of $G(k)$ k-th powers or less.*'

It is obvious that the existence of $g(k)$ involves that of $G(k)$, and that $G(k) \leqq g(k)$. When $k = 2$, both numbers are 4; but $G(3) \leqq 8$, by Landau's theorem, while $g(3) = 9$; and doubtless $G(k) < g(k)$ in general. It is important also to observe that, conversely, the existence of $G(k)$ involves that of $g(k)$. For, if $G(k)$ exists, all numbers beyond a certain limit γ are sums of $G(k)$ k-th powers or less. But all numbers less than γ are sums of γ ones or less, and therefore $g(k)$ certainly cannot exceed the greater of $G(k)$ and γ.

I said that $G(k)$ seemed to me the more fundamental of these numbers, and it is easy to see why. Let us assume (as is no doubt true) that the only numbers which require 9 cubes for their expression are 23 and 239. This is a very curious fact which should be interesting to any genuine arithmetician; for it ought to be true of an arithmetician that, as has been said of Mr. Ramanujan, and in his case at any rate with absolute truth, that 'every positive integer is one of his personal friends'. But it would be absurd to pretend that it is one of the profounder truths of higher arithmetic: it is nothing more than an entertaining arithmetical fluke. It is Landau's 8 and not Wieferich's 9 that is the profoundly interesting number.

The real value of $G(3)$ is still unknown. It cannot be less than 4; for every number is congruent to 0, or 1, or -1 to modulus 3, and it is an elementary deduction that every cube is congruent to 0, or 1, or -1 to modulus 9. From this it follows that the sum of three cubes cannot be of the form $9m + 4$ or $9m + 5$: for such numbers at least 4 cubes are necessary, so that $G(3) \geq 4$. But whether $G(3)$ is 4, 5, 6, 7, or 8 is one of the deepest mysteries of arithmetic.

It is worth while to glance at the evidence of computation. Dase, at the instance of Jacobi, tabulated the minimum number of cubes for values of n from 1 to $12,000$, and Daublensky von Sterneck has extended the table to $40,000$. Some of the results are shown in Table II. In each row I have shown a typical thousand numbers,

TABLE II.

	1	2	3	4	5	6	7	8	9
1– 1000	10	41	122	242	293	202	73	15	2
1000– 2000	2	27	113	283	358	194	23	—	—
9000–10000	1	17	121	377	401	83	—	—	—
19000–20000	1	12	100	400	426	61	—	—	—
29000–30000	1	11	105	448	388	47	—	—	—
39000–40000	1	13	117	457	384	28	—	—	—

classified according to the minimum number of cubes by which they can be expressed. There are 15 numbers only for which 8 are needed, the largest being 454. There are 121 for which 7 are needed, the two largest being 5818 and 8042; the distribution of these 121 numbers in the first 9 thousands is

$$73, \ 23, \ 7, \ 6, \ 7, \ 4, \ 0, \ 0, \ 1.$$

If empirical evidence means anything, it seems clear that $G(3) \leqq 6$. I am sure that Professor Townsend and Professor Lindemann have made countless generalizations on evidence far less substantial.

It is also clear that, throughout von Sterneck's tables, there is a fairly steady, though latterly very slow, decrease in the proportion of numbers for which even 6 cubes are required; but that the table is not sufficiently extensive to give any very decisive indication as to whether these numbers disappear or not. It seemed to me this was a case in which further evidence would be worth having. To calculate a *systematic* table on the scale required would be a work of years. It is possible, however, to obtain some indication of the probable truth, without any superhuman patience, by exploring a selected stratum of much larger numbers. Dr. Ruckle of Göttingen recently undertook this task at my request, and I am glad to be able to tell you his results. He found, for the 2,000 numbers immediately below 1,000,000, the following distribution.

	1	2	3	4	5	6	7
998000–999000	0	1	98	640	262	1	0
999000–1000000	1	1	94	614	289	1	0

You will observe that the 6-cube numbers have all but disappeared, and that there is a quite marked turnover from 5 to 4. Conjecture in such a matter is extremely rash, but I am on the whole disposed to predict with some confidence that $G(3) \leq 5$. If I were asked to choose between 5 and 4, all I could say would be this. That $G(3)$ should be 4 would harmonize admirably, so far as we can see at present, with the general trend of Mr. Littlewood's and my researches. But it is plain that, if the 5-cube numbers too do ultimately disappear, it can only be among numbers the writing of which would tax the resources of the decimal notation; and at present we cannot *prove* even that $G(3) \leq 7$, though here success seems not impossible.

With the fourth powers or biquadrates we have been very much more successful. I have explained that $g(4)$ lies between 19 and 37. As regards $G(4)$, we have here no numerical evidence on the same scale as for cubes. Any fourth power is congruent to 0 or 1 to modulus 16, and from this it follows that no number congruent to 15 to modulus 16 can be the sum of less than 15 fourth powers. Thus $G(4) \geq 15$; and Kempner, by a slight elaboration of this simple argument, has proved that $G(4) \geq 16$. No better upper bound was known before than the 37 of Wieferich, but here Mr. Littlewood and I have been able to make a very substantial improvement, first to 33 and finally to 21. Thus $G(4)$ lies between 16 and 21, and the margin is comparatively small.

I turn now to the general case. In the years up to 1909, the existence proof was effected, and upper bounds for $g(k)$ determined, for the values of k from 2 to 8 inclusive and for $k = 10$. These upper bounds are shown in the first row of Table III; that for 10, which is not included, is somewhere in the neighbourhood of $140,000$. In

TABLE III.

	2	3	4	5	6	7	8
$g(k) \leq$	4	9	37	58	478	3806	31353
$g(k) \geq \left[\left(\frac{3}{2}\right)^k\right] + 2^k - 2 =$	4	9	19	37	73	143	279
$G(k) \leq$	4	[8]	37	58	478	3806	31353
$G(k) \leq (k-2)2^{k-1} + 5 =$	(5)	(9)	**21**	**53**	**133**	**325**	**773**
$G(k) \geq k+1, 4k$	4	4	16	6	7	8	32

the second row I have shown the best known lower bounds, which are given by the simple general formula which stands to the left, in which $\left[\left(\frac{3}{2}\right)^k\right]$ denotes the integral part of $\left(\frac{3}{2}\right)^k$. It is easily verified, in fact, that the number

$$\left(\left[\left(\tfrac{3}{2}\right)^k\right] - 1\right) 2^k + 2^k - 1,$$

which is less than 3^k, requires the number of k-th powers stated.[1] It will be observed that the first three numbers are those which occur in Waring's enunciation.

Waring's problem, as I have defined it—the problem, that is to say, of finding a general existence proof for $g(k)$, and *a fortiori* for $G(k)$—was ultimately solved by Hilbert, once more in 1909. I wish that I had time to give a proper account of his justly famous memoir, which raised the whole discussion at once on to an altogether higher level. As it is, I must confine myself to one or two extremely inadequate remarks. The proof falls into two parts. The first part is what I may call semi-transcendental. It is not fully transcendental in the sense in which, for example, the classical proofs in the theory of the distribution of primes are transcendental, for it does not make use of the machinery of the theory of analytic functions of a complex variable; but it uses the methods of the integral calculus, and is therefore not fully elementary. Hilbert set out with what would appear at first sight to be the singularly ill-adapted weapon of a volume integral in space of 25 dimensions, a number which he was afterwards able to reduce to 5. The formula which he ultimately used is

$$(x_1^2 + x_2^2 + x_3^2 + x_4^2 + x_5^2)^k$$
$$= C \int\int\int\int\int (x_1 t_1 + x_2 t_2 + x_3 t_3 + x_4 t_4 + x_5 t_5)^{2k} \, dt_1 \ldots dt_5,$$

where C is a certain constant, viz.

$$\frac{(2k+1)(2k+3)(2k+5)}{8\pi^2},$$

---

[1] This observation was made by Bretschneider in 1853.

and the integration is effected over the interior of the hypersphere

$$t_1^2 + t_2^2 + t_3^2 + t_4^2 + t_5^2 = 1.$$

Starting from this formula he was able, by an exceedingly ingenious process based upon the definition of a definite integral as the limit of a finite sum, to prove the existence in the general case of algebraical identities analogous to that used by Liouville and his followers when k is 4. It should be observed that Hilbert's proof is essentially an *existence proof*; his method is not effective for the actual determination of these identities even in the simplest cases. The identities which are known for special values of k have been obtained by common algebra, and are, after the first few values of k, excessively complicated. The simplest known identity for $k = 10$, for instance, is

$$22680(x_1^2 + x_2^2 + x_3^2 + x_4^2)^5$$

$$= 9 \overset{(8)}{\sum}(x_1 \pm x_2 \pm x_3 \pm x_4)^{10} + \overset{(48)}{\sum}(2x_1 \pm x_2 \pm x_3)^{10}$$

$$+ 180 \overset{(12)}{\sum}(x_1 \pm x_2)^{10} + 9 \overset{(4)}{\sum}(2x_1)^{10},$$

where the figures in brackets show the number of terms under the signs of summation. However, the identities exist; and it should be clear to you, after our discussion of the case $k = 4$, that they enable us at once to obtain a proof in succession for $k = 2, 4, 8, 16, \ldots$ and generally whenever k is a power of 2. This concludes the first and most characteristic part of Hilbert's argument. The second part, in which the conclusion is extended to every value of k, is purely algebraical.

Hilbert's work has been reconsidered and simplified by a number of writers, most notably by Dr. Stridsberg of Stockholm, and the ultimate result of their work has been to eliminate the transcendental elements from the proof entirely. The proof, as they have left it, is fully elementary; it does not involve any reference to integrals, or to any kind of limiting process, but depends simply on an ingenious system of equations derived by the processes of finite algebra. It remains a pure existence proof, and throws no light on the value of $g(k)$.

It would hardly be possible for me to exaggerate the admiration which I feel for the solution of this historic problem of which I have been compelled to give so bald and summary a description. Within the limits which it has set for itself, it is absolutely and triumphantly successful, and it stands with the work of Hadamard and de la Vallée-Poussin, in the theory of primes, as one of the landmarks in the modern history of the theory of numbers. But there is an enormous amount which remains to be done, and it would seem that, if we are to interpret Waring's problem in the widest possible sense, if we are to get into real contact with the actual values of our numbers $g(k)$ and $G(k)$, still more if we are to attack all the obvious problems connected with the number of representations, then essentially different and inherently more powerful methods are required. There is one armoury only in which such more powerful weapons can be found, that of the modern theory of functions. In short we must learn how to apply Cauchy's Theorem to the problem, and that is what Mr. Littlewood and I have set out to do.

The first step is fairly obvious. The formulae are slightly simpler when k is *even*. The number of representations of n as the sum of s k-th powers, which we may denote in general by

$$r_{k,s}(n),$$

is then the coefficient of x^n in the generating function

$$1 + \sum_1^\infty r_{k,s}(n)x^n = (f(x))^s \,,$$

where

$$f(x) = 1 + 2x^{1^k} + 2x^{2^k} + 2x^{3^k} + \dots .$$

This formula involves certain conventions as to the order and sign of the numbers which occur in the representations which are to be reckoned as distinct; but the complications so introduced are trivial and I need not dwell on them. The series is convergent when $|x| < 1$, and, by Cauchy's Theorem, we have

$$r_{k,s}(n) = \frac{1}{2\pi i} \int \frac{(f(x))^s}{x^{n+1}} \, dx,$$

the path of integration being a circle whose centre is at the origin and whose radius is less than unity.

All this is simple enough; but the further study of the integral is very intricate and difficult, and I cannot attempt to do more than to give a rough idea of the obstacles that have to be surmounted. Let us contrast the integral for a moment with that which would stand in its place in the 'trivial' problem to which I referred early in my lecture. There the subject of integration would be a *rational* function, with a finite number of poles all situated on the unit circle. We could deform the contour into one which lies wholly at a considerable distance from the origin and in which, owing to the factor x^{n+1} in the denominator, every element is very small when n is large. We should have, of course, to introduce corrections corresponding to the residues at the poles; and it is just these corrections which would give the dominant terms of an approximate formula by means of which our coefficients could be studied. In the present case we have no such simple recourse;

for every point of the unit circle is a singularity of an exceedingly complicated kind, and the circle as a whole is a barrier across which it is impossible to deform the contour. It is of course for this reason that no successful application of the method has been made before.

Our fundamental idea for overcoming the difficulty is as follows. Among the continuous mass of singularities which covers up the circle, it is possible to pick out a class which to a certain extent dominates the rest. These special singularities are those associated with the *rational* points of the circle, that is to say, the points

$$x = e^{2p\pi i/q},$$

where p/q is a rational fraction in its lowest terms. This class of points is indeed an *infinite* class; but the infinity is, in Cantor's phrase, only an *enumerable* infinity; and the points can therefore be arranged in a simply infinite series, on the model of the series

$$\frac{0}{1}, \frac{1}{2}, \frac{1}{3}, \frac{2}{3}, \frac{1}{4}, \frac{3}{4}, \frac{1}{5}, \frac{2}{5}, \frac{3}{5}, \frac{4}{5}, \frac{1}{6}, \frac{5}{6}, \frac{1}{7}, \ldots$$

In the neighbourhood of these points the behaviour of the function is sufficiently complex indeed, but simpler than elsewhere. The function has, to put the matter in a rough and popular way, a general tendency to become large in the neighbourhood of the unit circle, but this tendency is most pronounced near these particular points. They are not only the *simplest* but also the *heaviest* singularities; their weight is greatest when the denominator q is smallest, decreases as q increases, and (as a physicist would say) becomes infinitely small when q is infinitely large. There is, therefore, at any rate, the hope that we may be able to isolate the contributions of each of these selected points, and obtain, by adding them together, a series which may give a genuine approximation to our coefficient.

I owe to Professor Harald Bohr of Copenhagen a picturesque illustration which may help to elucidate the general nature of our argument. Imagine the unit circle as a thin circular rail, to which are attached an infinite number of small lights of varying intensity, each illuminating a certain angle immediately in front of it. The brightest light is at $x = 1$, corresponding to $p = 0$, $q = 1$; the next brightest at $x = -1$, corresponding to $p = 1$, $q = 2$; the next at $x = e^{2\pi i/3}$ and $e^{4\pi i/3}$, and so on. We have to arrange the inner circle, the circle of integration, in the position of maximum illumination. If it is too far away the light will not reach it; if too near, the arcs which fall within the angles of illumination will be too narrow, and the light will not cover it completely. Is it possible to place it where it will receive a satisfactorily uniform illumination?

The answer is that this is *only* possible when k is 2. Our functions are then elliptic functions; the lights are the formulae of the theory of linear transformation; and we can find a position of the inner circle in which it falls entirely under their rays. We are thus led to a solution of the problem of the squares which is in all essential respects complete. But when k exceeds 2 the result is less satisfactory. The angle of the lights is then too narrow; the beams which they emit, instead of spreading out with reasonable regularity, are shaped like torpedoes or cigars; however we move our circle a part remains in darkness. It would seem that this difficulty, which held up our researches for something like two years, is the really characteristic difficulty of the general problem. It cannot be solved until we have found some other source of light.

It was only after the most prolonged and painful efforts that we were able to discover such another source. It is possible not only to hang lights upon the rail, but also, to a certain extent, to cause the rail itself to glow. The illumination which can be induced in this manner is irritatingly faint, and it is for this reason that our results are not yet all that we desire; but it is enough to make the dark places dimly visible and to enable us to prove a great deal more than has been proved before.

The actual results which we obtain are these. We find that there is a certain series, which we call the *singular series*, which is plainly the key to the solution. This series is

$$\mathbf{S} = \sum \left(\frac{S_{p,q}}{q} \right)^s e^{2np\pi i/q},$$

where

$$S_{p,q} = \sum_{h=0}^{q-1} e^{2h^k p\pi i/q}$$

—a sum which reduces, when $k = 2$, to one of what are known as 'Gauss's sums'—and the summation extends, first to all values of p less than and prime to q, and secondly to all positive integral values of q. The genesis of the series is this. We associate with the rational point $x = e^{2p\pi i/q}$ an auxiliary power series

$$f_{p,q}(x) = \sum_n c_{p,q,n} x^n,$$

which (a) is as simple and natural as we can make it, and (b) behaves perfectly regularly at all points of the unit circle except at the one point with which we are particularly concerned. We then add together all these auxiliary functions, and endeavour to approximate to the coefficient of our original series by summing the auxiliary coefficients over all values of p and q. The process is, at bottom, one of 'decomposition into simple elements', applied in an unusual way.

Our final formula for the number of representations is

$$r_{k,s}(n) = \frac{\left\{ 2\Gamma\left(1 + \frac{1}{k}\right) \right\}^s}{\Gamma\left(1 + \frac{s}{k}\right)} n^{\frac{s}{k}-1} \mathbf{S} + O(n^\sigma),$$

the second term denoting an error less than a constant multiple of n^σ, and σ being a number which is less than $\dfrac{s}{k} - 1$ at any rate for sufficiently large values of s. The second term is then of lower order than the first. Further, the first term is real, and it may be shown, if s surpasses a certain limit, to be *positive*. If both these conditions are satisfied, and n is sufficiently large, then $r_{k,s}(n)$ cannot be zero, and representations of n by s k-th powers certainly exist. The way is thus open to a proof of the existence of $G(k)$; if $G(k)$ exists, so also does $g(k)$, and Waring's problem is solved.

The structure of the dominant term in our general formula is best realized by considering some special cases. In Table IV I have written out the leading terms of $\mathbf{S}$, first when $k = 2$ and s is arbitrary, and then for 7 cubes and for 33 and 21 biquadrates. There are certain characteristics common to all these series. The terms diminish rapidly; in each case only a very few are of real importance: and they are oscillatory, with a period which increases as the amplitude of the oscillations decreases. The series for the cubes is easily shown to be positive; but we cannot deduce that $r_{3,7}(n)$ is positive, and draw consequences as to the representation of numbers by 7 cubes, because in this case we cannot dispose satisfactorily of the error term $O(n^\sigma)$ in the general formula. In the two cases relating to fourth powers which I have chosen, the discussion of the series itself is rather more delicate, for there is in each of them one term which can be negative and greater than 1. But the discussion can be brought to a satisfactory conclusion, and, as in this case we are able to prove that the error term is really of lower order, we obtain what we desire. *Every large number is the sum of* 21 *fourth powers or less*; $G(4) \leqq 21$. Further, we have obtained a genuine asymptotic formula for the number of representations, which can be used for the study of the representations of numbers of particular forms. We can show, for example, that a large number of the form $16n + 10$ can be

TABLE IV.

$$k = 2.$$

$$S = 1 + 0 + \frac{2}{3^{\frac{1}{2}s}} \cos\left(\frac{2}{3}n\pi - \frac{1}{2}s\pi\right) + \frac{2^{\frac{1}{2}s+1}}{4^{\frac{1}{2}s}} \cos\left(\frac{1}{2}n\pi - \frac{1}{4}s\pi\right)$$

$$+ \frac{2}{5^{\frac{1}{2}s}}\left\{\cos\frac{2}{5}n\pi + \cos\left(\frac{4}{5}n\pi - s\pi\right)\right\} + 0 + \ldots.$$

$$k = 3, s = 7.$$

$$S = 1 + 0.610 \cos\tfrac{2}{9}n\pi + 0.130 \cos\tfrac{2}{7}n\pi + 0.078 \cos\tfrac{6}{7}n\pi + \ldots.$$

$$k = 4, s = 33.$$

$$S = 1 + 1.054 \cos(\tfrac{1}{8}n\pi - \tfrac{1}{16}\pi) + 0.147 \cos(\tfrac{1}{4}n\pi - \tfrac{1}{8}\pi) + \ldots.$$

$$k = 4, s = 21.$$

$$S = 1 + 1.331 \cos(\tfrac{1}{8}n\pi + \tfrac{11}{16}\pi) + 0.379 \cos(\tfrac{1}{4}n\pi - \tfrac{5}{8}\pi) + \ldots.$$

expressed by 21 biquadrates in about 200 times more ways than one of the form $16n + 2$.

If the method of which I have tried to give some general idea is compared with those which have previously been applied to the problem, it will be found that it has three very great advantages. In the first place it is inherently very much more powerful. It brings us for the first time into relation with the series on which the solution in the last resort depends, and tells us, approximately but truly, what the number of representations really is. Secondly, it gives us numerical results which, as soon as k exceeds 3, are far in advance of any known before. These numbers are those in the fourth row of Table III.[2] It will be seen that these numbers conform to a simple law, and that is the third advantage of the method, that it is not a mere existence proof, but gives us a definite upper bound for $G(k)$ for all values of k, viz.

$$G(k) \leqq (k-2)2^{k-1} + 5.$$

In the last row of the table I have shown all that is known about $G(k)$ on the other side. In all cases $G(k) \geqq k + 1$, while if k is a power of 2 we can say more, namely that $G(k) \geqq 4k$. A comparison between this row of figures and that above it is enough to show the room which remains for further research. It is beyond question that our numbers are still very much too large; and there is no sort of finality about our researches, for which the best that we claim is that they embody a method which opens the door for more.

I will conclude by one word as to the application of our method to another and a still more difficult problem. It was asserted by Goldbach in 1742 that *every even number is the sum of two odd primes.* Goldbach's assertion remains unproved; it has not even been proved that every number n is the sum of 10 primes, or of 100, or of any number independent of n. Our method is applicable in principle to this problem also. We cannot solve the problem, but we can open the first serious attack upon it, and bring it into relation with the established prime number theory. The most which we can accomplish at present is as follows. We have to assume the truth of the notorious Riemann hypothesis concerning the zeros of the Zeta-function, and

---

[2]The thick type indicates a new result. The (5) and (9) in round brackets are inferior to results already known. Our method is easily adapted to deal with the case $k = 2$ completely; but it will not at present yield Landau's 8, which is therefore enclosed in square brackets.

indeed in a generalized and extended form. If we do this we can prove, not Goldbach's Theorem indeed, but the next best theorem of the kind, viz. that *every odd number*, at any rate from a certain point onwards, *is the sum of three odd primes*. It is an imperfect and provisional result, but it is the first serious contribution to the solution of the problem.

POSTSCRIPT

Srinivasa Ramanujan, F.R.S., Fellow of Trinity College, Cambridge, died in India on April 26, 1920, aged 32.

An account of his life and mathematical activities will be published in Vol. 19 of the *Proceedings of the London Mathematical Society.*

33186431R00023

Printed in Poland
by Amazon Fulfillment
Poland Sp. z o.o., Wrocław